Sun-Day Green

AF715331

Sun-Day Green

Pushpendra Gupta

Copyright © 2026 by Pushpendra Gupta
All rights reserved. No part of this book may be reproduced in any manner whatsoever without written permission except in the case of brief quotations embodied in critical articles and reviews.
First Printing, 2026

CONTENTS

1 Introduction

The sun has always been central to Indian culture, revered as a life-giving deity in ancient texts and traditions. Today, this celestial body offers more than spiritual significance—it presents a practical solution to one of India's most pressing challenges: energy security in a sustainable world.

India stands at a critical juncture. With a population of over 1.4 billion people and an economy growing at a remarkable pace, the country's energy demands continue to surge. Yet this growth comes with complications. For decades, India has relied heavily on coal and imported oil to power its development, creating a dual burden of environmental degradation and economic vulnerability.

The numbers tell a compelling story. India is the world's third-largest energy consumer, yet hundreds of millions of citizens still face unreliable access to electricity. The country imports over 80% of its oil needs, making it vulnerable to global price fluctuations and geopolitical tensions. Meanwhile, air quality in major cities ranks among the worst globally, with fossil fuel combustion being a primary contributor.

Solar energy emerges as a powerful answer to these challenges. India receives an average of 300 sunny days per year across most regions, with solar radiation intensity that exceeds many countries that have already embraced solar technology. This abundant natural resource remains largely untapped, representing not just an environmental opportunity but an economic one.

This book, "Sunday Green – India ka Solar Energy," aims to illuminate the path toward India's solar transformation. We'll explore how solar power can address the country's energy challenges while creating jobs, improving public health, and helping meet international climate commitments. From technical foundations to policy frameworks, from urban rooftops to rural microgrids, we'll examine how solar energy is reshaping India's energy landscape.

The following pages offer more than just information—they provide a vision of possibility. For policymakers, entrepreneurs, students, and concerned citizens alike, understanding India's solar potential is the first step toward harnessing it. As we'll see, the solar revolution isn't just coming to India—it's already begun.

2

Solar Revolution in India

Energy Landscape Evolution

The story of India's energy begins long before independence, shaped by colonial priorities and post-independence development goals. During British rule, energy infrastructure primarily served colonial interests—powering railways that transported raw materials, illuminating administrative centers, and energizing industries that processed India's resources for export. The first electricity generation station in India was established in 1897 at Darjeeling, followed by hydroelectric power stations in the early 1900s in Karnataka and Tamil Nadu.

Post-independence India faced the monumental task of building an energy system that could support national development. The 1950s and 1960s saw the establishment of state electricity boards and the creation of massive hydroelectric projects like the Bhakra Nangal Dam. Prime Minister Jawaharlal Nehru famously called these dams the "temples of modern India," reflecting the nation's faith in large-scale infrastructure as the path to progress.

The 1970s marked a turning point with the global oil crisis exposing India's vulnerability to international energy markets. This period prompted the first serious considerations of energy security and self-sufficiency. The government established the Department of Non-Conventional Energy Sources in 1982 (later elevated to the Ministry of New and Renewable Energy), signaling early recognition of alternative energy's importance.

Coal emerged as the backbone of India's energy strategy through the 1980s and 1990s. The country's abundant coal reserves made it an economically attractive option, leading to massive expansion of coal-fired power plants. By the early 2000s, coal accounted for over 70% of India's electricity generation, cementing a dependency that continues to this day.

The liberalization of India's economy in the 1990s brought private sector participation in energy generation, transmission, and distribution. This period saw increased investment in power infrastructure, though primarily in conventional energy sources. The Electricity Act of 2003 further restructured the sector, creating a more competitive market environment and laying groundwork for future renewable energy development.

Throughout this evolution, a troubling pattern emerged: despite enormous growth in generation capacity, energy access remained uneven. Urban areas received priority while rural India lagged behind. By the early 2000s, nearly 400 million Indians still lacked reliable electricity access, highlighting the limitations of the centralized fossil fuel-based approach.

The environmental consequences of this development path became increasingly apparent. Major cities like Delhi, Mumbai, and Kolkata began experiencing severe air pollution, with power plants and industrial emissions as major contributors. India's carbon footprint grew substantially, placing the country among the world's largest greenhouse gas emitters, though its per capita emissions remained far below developed nations.

The 2010s marked the beginning of a new chapter in India's energy story. Rising concerns about climate change, air quality, and energy security converged with rapidly falling costs of renewable technologies. Solar power, once considered prohibitively expensive, began to compete with conventional sources on price alone. This economic shift, combined with growing environmental awareness, created fertile ground for India's solar revolution.

In 2010, the Jawaharlal Nehru National Solar Mission launched with an initial target of 20 gigawatts (GW) of solar capacity by 2022. This modest beginning would soon transform into one of the world's most ambitious renewable energy programs, reflecting a fundamental shift in India's approach to energy development.

The evolution of India's energy landscape reveals a nation grappling with competing priorities: development needs, resource constraints, environmental concerns, and technological opportunities. The emergence of solar energy represents not just a new technology but a potential resolution to these tensions—a path that aligns economic growth with environmental sustainability and energy security.

National Solar Mission Blueprint

The National Solar Mission, launched in 2010 and later renamed the Jawaharlal Nehru National Solar Mission (JNNSM), represents India's most comprehensive policy framework for solar energy development. What began as a targeted initiative has evolved into a cornerstone of India's renewable energy strategy and climate action plan.

The mission's original goals were modest by today's standards: 20 GW of grid-connected solar power by 2022, divided into three phases. Phase I (2010-2013) focused on capturing the low-hanging fruit through government-supported projects. Phase II (2013-2017) aimed to scale up capacity and create conditions for market competitiveness. Phase III (2017-2022) envisioned achieving grid parity—the point where solar electricity costs the same as conventional sources without subsidies.

In 2015, Prime Minister Narendra Modi's government dramatically expanded these targets, announcing a new goal of 100 GW of solar capacity by 2022—a fivefold increase from the original plan. This revised target included 60 GW from utility-scale projects and 40 GW from rooftop installations. This ambitious revision positioned India among global leaders in renewable energy commitment and signaled the country's seriousness about solar power as a mainstream energy source.

The implementation framework of the National Solar Mission operates through multiple mechanisms designed to address different market segments and barriers. For utility-scale projects, the government introduced competitive bidding processes where developers bid to supply solar power at the lowest possible tariff. This approach has driven remarkable price discoveries, with solar tariffs falling from over ₹17 per kilowatt-hour in 2010 to under ₹2 in recent auctions—cheaper than most coal-fired power.

For rooftop solar, the mission employs a combination of capital subsidies, accelerated depreciation benefits for commercial installations, and net metering policies that allow consumers to sell excess power back to the grid. The government has also established dedicated solar parks—large tracts of land with necessary infrastructure where multiple developers can set up projects, reducing development risks and costs.

Financing mechanisms form another crucial component of the mission. The Indian Renewable Energy Development Agency (IREDA) provides specialized loans for renewable projects. International funding has been mobilized through partnerships with the World Bank, Asian Development Bank, and other institutions. Green bonds have emerged as an innovative financing tool, with both public and private entities raising capital specifically for clean energy projects.

The policy architecture extends beyond direct support for solar installations. Manufacturing incentives aim to build domestic production capacity for solar equipment, reducing dependence on imports (particularly from China) and creating local jobs. Research and development initiatives focus on adapting solar technologies to Indian conditions and developing innovative applications like solar pumps for agriculture and floating solar plants on reservoirs.

Implementation of the National Solar Mission has not been without challenges. The rooftop segment has consistently underperformed against targets due to regulatory hurdles, financing difficulties, and consumer awareness gaps. Land acquisition for large-scale projects remains complicated in a densely populated country with competing land uses. Grid integration presents technical challenges as the share of variable renewable energy increases. Domestic manufacturing has struggled to compete with cheaper imports despite policy support.

Despite these obstacles, the mission has achieved remarkable progress. From less than 10 MW of solar capacity in 2010, India reached approximately 50 GW by early 2022. While this fell short of the 100 GW target, it still represents one of the fastest expansions of solar power anywhere in the world. More importantly, the mission has transformed solar energy from a niche alternative to a mainstream power source that regularly outcompetes fossil fuels on cost alone.

The National Solar Mission has evolved beyond its original conception as an energy program to become a multidimensional initiative with economic, environmental, and social dimensions. It addresses energy access through off-grid solar applications in remote areas. It contributes to skill development and job creation across the solar value chain. It forms a central pillar of India's climate action strategy, supporting the country's commitment under the Paris Agreement to achieve 40% of installed electric capacity from non-fossil sources by 2030.

Looking ahead, the mission continues to adapt to changing circumstances. New targets aim for 280 GW of solar capacity by 2030 as part of a larger 500 GW renewable energy goal. Policy innovations like solar-wind hybrid projects, battery storage integration, and green hydrogen production are expanding the mission's scope. What began as a targeted solar program has grown into a comprehensive framework for energy transition that touches virtually every aspect of India's economy and society.

Solar Potential versus Traditional Sources

When comparing solar energy to conventional power sources in India, we must consider multiple dimensions: resource availability, economic factors, technical characteristics, environmental impact, and social implications. This multifaceted analysis reveals why solar power has moved from alternative to mainstream in India's energy planning.

India's solar resource is exceptional by global standards. The country receives an average annual solar radiation of 4-7 kWh per square meter per day across most regions, with over 300 sunny days in many areas. This translates to a theoretical potential of about 750 GW, assuming just 3% of wasteland area is used for solar installations. When rooftops, agricultural land (through dual-use approaches), and innovative applications like canal-top solar are included, the potential exceeds 1,000 GW—far surpassing India's current total installed electricity capacity of approximately 400 GW.

In contrast, India's conventional energy resources face limitations. Coal reserves, while substantial at around 100 billion tons, are finite and concentrated in specific regions, creating logistical challenges and regional disparities. Natural gas reserves are limited, forcing heavy import dependence. Hydropower potential, estimated at 145 GW, faces increasing constraints from environmental concerns, displacement issues, and climate change impacts on water availability.

The economic comparison between solar and conventional sources has undergone a dramatic shift. In 2010, solar power in India cost approximately ₹17 per kilowatt-hour, making it among the most expensive generation options. By 2021, utility-scale solar tariffs had fallen below ₹2 per kilowatt-hour in competitive auctions—cheaper than new coal plants and even below the operating cost of some existing thermal power stations. This price revolution resulted from global technology improvements, economies of scale, and India's competitive policy framework.

Capital costs tell a similar story. While solar requires higher upfront investment per megawatt than coal, it has no fuel costs and minimal operational expenses. The levelized cost of electricity (LCOE)—which accounts for all lifetime costs—now favors solar in most scenarios. Additionally, solar projects can be built much faster (typically 6-12 months) compared to coal plants (4-5 years) or nuclear facilities (10+ years), allowing quicker response to energy needs.

From a technical perspective, both solar and conventional sources have distinct characteristics. Solar power's variability—it generates only during daylight hours and is affected by weather conditions—presents integration challenges for the grid. However, this limitation is increasingly addressed through geographic distribution of installations, hybrid projects combining solar with wind or hydro, and emerging storage technologies. India's diverse geography actually helps mitigate variability, as cloud cover or monsoon conditions affecting one region often leave others unaffected.

Conventional power plants offer advantages in dispatchability (ability to generate on demand) and inertia (which helps maintain grid stability). However, they face their own technical challenges, including decreasing efficiency in India's water-stressed and increasingly hot climate. Coal plants designed to run continuously struggle in a system requiring more flexibility to accommodate growing renewable generation.

The environmental comparison strongly favors solar energy. A typical coal plant in India emits approximately 1 kg of CO2 per kilowatt-hour, along with sulfur dioxide, nitrogen oxides, particulate matter, and mercury. These emissions contribute to climate change and cause severe local air pollution—a major public health crisis in India. Solar power produces zero emissions during operation, with life-cycle emissions (from manufacturing and construction) estimated at just 20-50 grams of CO2 per kilowatt-hour.

Water usage presents another stark contrast. Thermal power plants require 3-5 cubic meters of water per megawatt-hour for cooling, making them major consumers in a water-stressed country. Solar photovoltaic systems need only occasional cleaning, using approximately 0.1 cubic meters per megawatt-hour—a 97% reduction. This water advantage becomes increasingly valuable as climate change exacerbates water scarcity across India.

Land use considerations are more complex. Utility-scale solar requires approximately 5-6 acres per megawatt, seemingly more than coal plants. However, this comparison misleads without considering the land disturbed by coal mining, ash disposal, and transportation infrastructure. Solar installations can also utilize otherwise unproductive land, be integrated into existing structures (rooftops), or combined with agriculture through agrivoltaics.

From a social and economic development perspective, solar energy offers unique advantages. The distributed nature of solar resources allows for more geographically balanced development compared to conventional energy, which concentrates around fuel sources or major transmission corridors. Solar creates more jobs per unit of energy—approximately 3.45 job-years per gigawatt-hour compared to 0.27 for coal, according to some studies. These jobs span various skill levels, from manufacturing and installation to operations and maintenance.

Energy security represents another dimension where solar outperforms traditional sources. India imports over 80% of its oil, 50% of its natural gas, and increasingly even coal, creating economic vulnerability and geopolitical dependence. Solar power utilizes an indigenous resource, reducing import bills and insulating the economy from international fuel price volatility. During the 2020 COVID-19 pandemic and subsequent global energy crisis, renewable energy proved more resilient than fossil fuel supply chains.

The comparison between solar and conventional energy is not simply about replacing one technology with another—it involves reimagining the entire energy system. Traditional energy developed around a centralized model with large generation plants, extensive transmission networks, and one-way power flow. Solar enables a more distributed, flexible, and participatory system where consumers can also become producers through rooftop installations.

This transformation aligns with broader technological trends toward digitalization, decentralization, and democratization of infrastructure. Smart meters, battery storage, electric vehicles, and digital platforms for energy trading all complement solar technology, creating opportunities for innovation beyond what conventional energy systems can offer.

The evidence increasingly suggests that solar energy is not merely an alternative to traditional sources but a superior option for India's specific context—abundant in resource, increasingly cost-effective, environmentally necessary, and aligned with the country's development needs. The question has shifted from whether solar can compete with conventional energy to how quickly the transition can occur and how to manage the integration challenges along the way.

speed around the Sun, the time shown on a sundial can differ from true „clock" time. The correction needed to convert sundial time to clock time is called the Equation of Time.

$c = (h^2 + r^2)^{1/2}$

$h/r = \tan(\Theta)$

$\Theta = \tan^{-1}(h/r)$

Where

h = height of gnomon

r = length of shadow

Θ = incident angle of light source (sun or artificial light source)

Fig 6

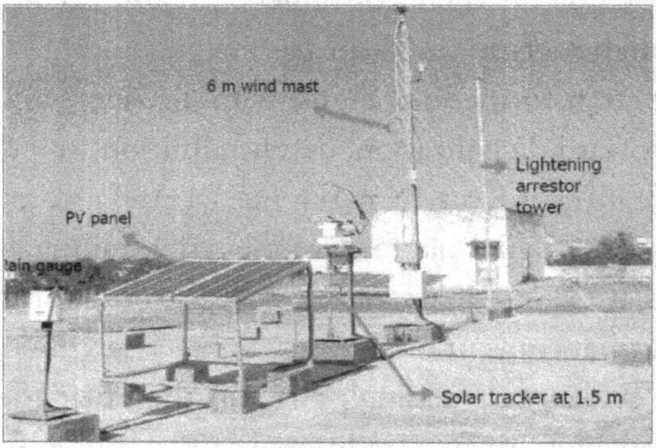

Fig 7

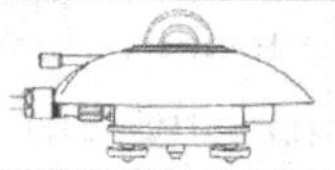

Fig 8

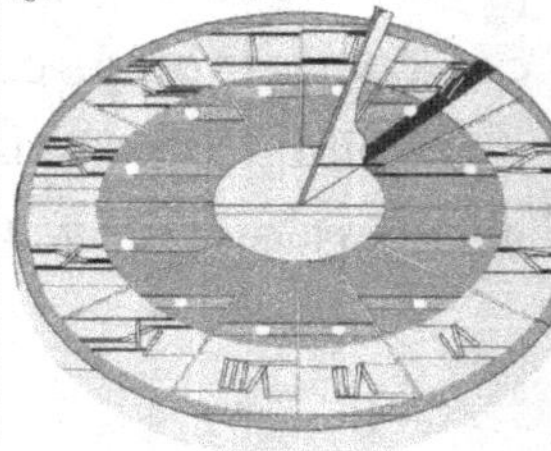

Sunlight spectrum (Fig 9)

Fig 9

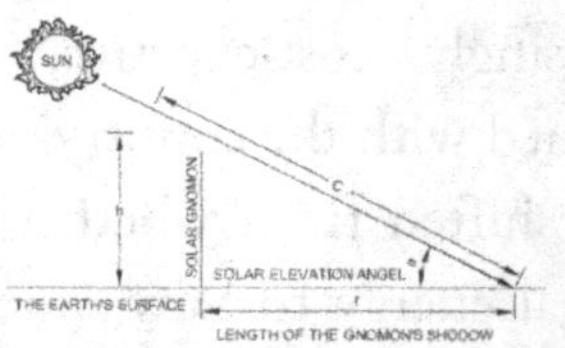

The sun's outermost and relatively thin 400 km layer is called the Photosphere and has a temperature of approximately 5,770 Kelvin. This is the layer that emits the spectrum of radiation which is visible to the human eye and is termed 'light'.

Scattering of solar radiation takes place within the whole spectral range. However, there are different ways in which the scattering can occur:

3

Technical Foundations of Solar Power

Solar Panel Technology Landscape

The foundation of India's solar revolution rests on photovoltaic (PV) technology—the direct conversion of sunlight into electricity through semiconductor materials. Understanding the various solar panel technologies, their performance characteristics, and suitability for Indian conditions is essential for stakeholders across the solar ecosystem.

Three main technologies dominate India's solar landscape: monocrystalline silicon, polycrystalline silicon, and thin-film. Each offers distinct advantages and limitations that determine their application in different contexts.

Monocrystalline silicon panels represent the premium segment of the market. Manufactured from single-crystal silicon ingots, these panels feature a distinctive uniform black appearance and rounded edges. Their key advantage lies in efficiency—typically 18-22% in commercial products, with laboratory versions exceeding 26%. This higher efficiency means more power generation per square meter, making them ideal for space-constrained installations like urban rooftops.

The manufacturing process for monocrystalline panels involves growing single-crystal silicon ingots using the Czochralski method, slicing these into wafers, doping them to create p-n junctions, adding electrical contacts, and assembling them into modules with protective glass, encapsulant, and backsheet. This complex process results in higher costs but produces panels with superior performance and longevity.

In Indian conditions, monocrystalline panels offer several advantages. Their lower temperature coefficient (typically -0.35% to -0.45% per degree Celsius above 25°C) means they maintain better performance during hot weather compared to alternatives. This characteristic proves valuable in regions like Rajasthan, Gujarat, and central India, where summer temperatures regularly exceed 40°C. Their higher efficiency also means fewer panels and less supporting infrastructure for a given capacity, potentially offsetting the higher upfront cost.

Polycrystalline (or multicrystalline) silicon panels constitute the largest segment of India's installed solar capacity. These panels are manufactured by melting silicon fragments and casting them into blocks, resulting in a distinctive blue, multifaceted appearance. Commercial efficiencies typically range from 15-18%, lower than monocrystalline but with a more favorable cost-to-performance ratio that has driven widespread adoption.

The manufacturing process for polycrystalline panels is less energy-intensive and wastes less silicon, resulting in lower production costs. The molten silicon is cast into blocks, which are then cut into wafers, processed similarly to monocrystalline cells, and assembled into modules. This simpler process has made polycrystalline technology particularly attractive for large-scale manufacturing, including in India's domestic production facilities.

In Indian operating conditions, polycrystalline panels show slightly higher performance degradation in extreme heat, with temperature coefficients typically between -0.40% and -0.50% per degree Celsius. However, they often perform better in diffuse light conditions, such as during hazy days or the monsoon season, partially compensating for this limitation. Their cost advantage has made them the preferred choice for utility-scale installations where land constraints are less severe.

Thin-film technologies represent a smaller but significant segment of India's solar market. Unlike crystalline silicon panels, thin-film modules deposit thin layers of photovoltaic material (such as cadmium telluride, copper indium gallium selenide, or amorphous silicon) onto glass, plastic, or metal substrates. These technologies typically achieve lower efficiencies (10-16% commercially) but offer unique advantages in specific applications.

The manufacturing process for thin-film involves depositing semiconductor materials in thin layers (typically a few micrometers thick, compared to 200+ micrometers for silicon wafers) onto substrates using various deposition techniques. This process uses less material and can be more automated than crystalline silicon production, though economies of scale have been harder to achieve due to lower market share.

In Indian conditions, thin-film technologies demonstrate some distinct advantages. They typically have better temperature coefficients (-0.25% to -0.35% per degree Celsius) than crystalline silicon, maintaining more of their rated output in hot weather. They also perform better in low-light and diffuse light conditions, making them suitable for areas with frequent haze or dust. Their uniform appearance has made them popular for building-integrated applications where aesthetics matter.

Beyond these mainstream technologies, several emerging options show promise for India's solar future. Bifacial modules, which can generate electricity from both sides by capturing reflected light, offer 5-30% additional yield depending on installation conditions. These have gained traction in desert regions of Rajasthan and Gujarat, where high ground reflectivity maximizes the bifacial gain.

Heterojunction technology (HJT), combining crystalline silicon with thin-film amorphous silicon layers, achieves higher efficiencies while maintaining better temperature performance. Though currently more expensive, domestic manufacturing initiatives like Reliance's gigafactory plans include HJT production, potentially making this technology more accessible in the Indian market.

Perovskite solar cells, still primarily in the research phase, offer the prospect of low-cost, high-efficiency panels that could be manufactured using simpler processes. Indian research institutions like the Indian Institute of Science (IISc) and National Chemical Laboratory are actively working on perovskite technology adapted to Indian conditions.

The performance of these various technologies in India's diverse climate zones reveals important patterns. In the hot, arid regions of western India, the temperature coefficient becomes a critical factor, favoring monocrystalline and thin-film options despite their different price points. In the humid coastal regions, resistance to corrosion and salt mist becomes essential for longevity. In dusty environments, which characterize many of India's high-solar-potential regions, panel design features like frameless construction or self-cleaning glass can maintain better performance between cleanings.

India's unique challenges have driven innovation in panel design specifically for local conditions. These include:

1. Enhanced heat dissipation features to maintain performance during extreme summer temperatures
2. Reinforced construction to withstand high wind loads during monsoon and cyclone seasons
3. Improved dust resistance and self-cleaning properties for arid regions
4. Specialized glass treatments to reduce soiling in agricultural areas
5. Corrosion-resistant components for coastal installations

The domestic manufacturing landscape for solar panels in India has evolved significantly, but faces persistent challenges. From less than 500 MW of annual production capacity in

2014, domestic manufacturing has grown to approximately 10 GW for modules and 3 GW for cells by 2022. However, this remains insufficient to meet India's installation targets, resulting in continued import dependence, primarily on China.

Government initiatives to boost domestic manufacturing include production-linked incentive schemes, mandatory domestic content requirements for certain projects, and import duties on foreign panels. These measures aim to build manufacturing capacity across the value chain—from polysilicon and wafers to cells and modules—creating a self-reliant solar ecosystem.

Quality considerations remain paramount regardless of technology choice. India's Ministry of New and Renewable Energy (MNRE) has established the Solar Photovoltaics, Systems, Devices and Components Goods (Requirements for Compulsory Registration) Order, making BIS certification mandatory for solar modules. This certification ensures panels meet minimum quality standards for use in Indian conditions, addressing earlier issues with substandard imports.

The International Electrotechnical Commission (IEC) standards provide additional quality benchmarks, with IEC 61215 for design qualification and IEC 61730 for safety being particularly relevant. Premium installations often specify additional requirements like enhanced mechanical load testing, potential-induced degradation (PID) resistance, and extended reliability testing.

For project developers and consumers, technology selection involves balancing multiple factors: upfront cost, space constraints, expected lifetime, degradation rates, and specific site conditions. While polycrystalline technology dominated India's early solar growth due to cost advantages, monocrystalline panels have gained market share as prices have fallen and efficiency has become more valued. Thin-film technologies maintain a niche in specific applications where their unique properties offer advantages.

The technology landscape continues to evolve rapidly. Innovations like larger wafer sizes (from the traditional 156mm to 210mm), half-cut cells that reduce resistive losses, multi-busbar designs that improve current collection, and advanced encapsulation materials are continuously improving performance while reducing costs. These incremental improvements, combined with manufacturing scale, have driven the remarkable cost reductions that make solar increasingly competitive with conventional energy sources.

As India pursues its ambitious solar targets, the technology choices made today will shape the performance, reliability, and economics of installations that will generate power for decades to come. Understanding the nuances of different panel technologies and their suitability for India's diverse conditions is therefore essential for maximizing the country's solar potential.

Grid Integration Strategies

The integration of solar energy into India's electrical grid represents both a technical challenge and a strategic opportunity. As solar capacity grows from a few percentage points to a substantial portion of the energy mix, the approach to grid integration must evolve from simple accommodation to fundamental redesign. This transformation spans both grid-connected and off-grid applications, each with distinct requirements and solutions.

Grid-connected solar systems in India fall into two broad categories: utility-scale plants and distributed rooftop installations. Utility-scale plants, typically ranging from 1 MW to over 2,000 MW, connect directly to transmission networks and function essentially as power plants. Rooftop systems, usually between 1 kW and 500 kW, connect to distribution networks and serve local loads while potentially exporting surplus power.

For utility-scale plants, grid integration begins with interconnection infrastructure. These systems require substations with transformers to step up the voltage from the plant's internal collection system (typically 33 kV) to transmission levels (132/220/400 kV). Protection equipment, including circuit breakers and relays, must be coordinated with grid requirements to ensure safe operation. Advanced metering infrastructure tracks energy production and enables billing under power purchase agreements.

The technical requirements for utility-scale solar plants have become increasingly stringent as their grid presence grows. Early plants were simply required to deliver energy, but newer installations must provide grid support functions. These include:

1. Voltage control through reactive power management
2. Frequency response capabilities to help stabilize the grid

3. Ramp rate controls to prevent sudden output changes
4. Fault ride-through capability to remain connected during grid disturbances
5. Communication interfaces with grid operators for visibility and control

The Central Electricity Authority's "Technical Standards for Connectivity to the Grid" (Amendment) Regulations, 2019, formalized many of these requirements, marking India's transition toward treating solar as a mainstream generation resource with responsibilities beyond simply delivering energy.

Forecasting and scheduling represent another critical aspect of utility-scale integration. Unlike conventional plants that can be dispatched on demand, solar generation varies with weather conditions. The Indian Energy Exchange and power system operators now require solar generators to provide day-ahead and intra-day forecasts of expected generation. Deviations beyond permitted margins incur financial penalties, creating incentives for accurate forecasting.

Several forecasting approaches have emerged to meet this need. Physical models use weather prediction data and plant characteristics to estimate output. Statistical models analyze historical production patterns and weather correlations. Hybrid approaches combine these methods, often incorporating machine learning algorithms to improve accuracy. Specialized service providers now offer forecasting solutions tailored to Indian conditions, helping plant operators minimize deviation penalties.

For distributed rooftop solar, integration challenges differ substantially. These systems connect to low or medium-voltage distribution networks not originally designed for two-way power flow. Technical issues include potential voltage fluctuations, phase imbalances, and protection coordination problems when solar generation exceeds local demand.

Net metering policies have been the primary enabler for rooftop solar integration in India. These arrangements allow consumers to export surplus power to the grid, effectively using the network as storage and offsetting consumption during non-generating hours. Most Indian states have implemented net metering regulations, though with varying capacity limits, compensation mechanisms, and technical requirements.

The implementation of net metering has faced practical challenges. Distribution companies (DISCOMs) often lack experience with small-scale generators, resulting in complicated approval processes and connection delays. Technical standards for inverters and protection equipment have evolved unevenly across states. Billing systems designed for one-way consumption have required upgrades to handle bidirectional energy flows.

More advanced approaches to distributed solar integration are emerging. Gross metering separates consumption and generation, allowing different tariffs for each. Virtual net metering enables shared solar facilities to offset consumption at multiple locations. Time-of-day pricing creates incentives to align consumption with generation or to incorporate storage. These mechanisms represent a gradual evolution toward more sophisticated integration of distributed resources.

Off-grid solar systems serve areas beyond the reach of conventional electricity networks or provide backup during grid outages. These systems range from simple solar lanterns and home systems to complex microgrids serving entire communities. While not directly connected to the main grid, they form an important part of India's solar ecosystem and face their own integration challenges.

Solar home systems typically include panels, batteries, charge controllers, and inverters sized for basic household needs. The integration challenge here lies in component compatibility, user interfaces, and maintenance requirements. Standardization efforts like the Global LEAP Awards have helped establish quality benchmarks, while pay-as-you-go business models have made these systems more accessible by addressing upfront cost barriers.

Solar microgrids represent a more sophisticated off-grid approach, serving clusters of homes, businesses, and community facilities with a shared generation and distribution system. These installations require careful load assessment, generation sizing, and control systems to maintain stable operation. Many incorporate hybrid generation (solar plus diesel or biomass) to ensure reliability during extended low-solar periods.

The technical architecture of microgrids has evolved substantially. Early systems used simple charge controllers and basic lead-acid batteries. Modern installations incorporate advanced battery management systems, sophisticated inverters with frequency and voltage control capabilities, and energy management systems that optimize generation, storage, and load. Some are designed as "grid-ready," capable of interconnecting with the main grid if it eventually reaches their location.

The integration of storage technologies represents a frontier in both grid-connected and off-grid solar applications. Batteries serve different functions depending on the context: in off-grid systems, they provide essential night-time power; in rooftop installations, they increase self-consumption; in utility-scale plants, they can provide dispatchability and grid services.

Battery technologies deployed in Indian solar applications include:

1. **Lead-acid batteries:** Still common in budget-sensitive applications despite limited cycle life
2. **Lithium-ion batteries:** Increasingly dominant due to falling costs and superior performance
3. **Flow batteries:** Emerging in larger installations where their scalability and long-duration capabilities offer advantages
4. **Sodium-based batteries:** Gaining interest as a potentially lower-cost alternative to lithium for stationary applications

The integration of these storage technologies requires specialized battery management systems, thermal management (particularly important in India's hot climate), and control software that optimizes charging and discharging based on generation patterns, load requirements, and battery characteristics.

Beyond batteries, other integration technologies are gaining traction. Smart inverters with grid-supportive functions can help maintain stable voltage and frequency. Energy management systems optimize the interaction between solar generation, storage, and loads. Demand response capabilities adjust consumption patterns to better match variable generation. Advanced forecasting tools improve operational planning and grid coordination.

The regulatory framework for solar grid integration continues to evolve. The Electricity Act amendments, forecasting and scheduling regulations, technical standards for connectivity, and various state-level policies create a complex landscape that solar developers must navigate. This framework increasingly recognizes solar not just as an alternative energy source but as a fundamental component of the future grid.

Looking ahead, India's grid integration approach is moving toward a more holistic system perspective. The Green Energy Corridors project, dedicated transmission infrastructure for renewable energy, represents one aspect of this evolution. Proposals for renewable energy management centers, enhanced forecasting systems, and market mechanisms for flexibility services reflect growing recognition that integrating high levels of solar requires systemic changes rather than incremental adaptations.

The ultimate vision is a flexible, resilient grid where solar and other renewable sources form the foundation of the energy system rather than exceptions accommodated within a conventional framework. Achieving this vision will require continued technical innovation, regulatory evolution, and institutional capacity building across India's diverse electricity ecosystem.

Solar Performance Optimization

Maximizing the return on solar investments in India requires systematic approaches to performance optimization across the system lifecycle. From initial design through decades of operation, each phase presents opportunities to enhance energy yield, extend equipment life, and improve financial returns. This multifaceted optimization process combines technical expertise, data analytics, and economic assessment.

Performance optimization begins during system design. Site-specific factors dramatically influence energy production, making thorough assessment essential. Solar resource assessment typically starts with satellite-derived irradiation data from sources like the National Solar Radiation Database or SolarGIS, providing historical patterns of direct and diffuse radiation. For larger projects, this data is often supplemented with ground measurements using pyranometers and other instruments to validate satellite estimates and capture local microclimatic conditions.

Beyond raw resource availability, design optimization must account for India's diverse environmental challenges. In the northern plains, seasonal dust storms can reduce panel output by 15-30% without proper mitigation strategies. In coastal regions like Gujarat and Tamil Nadu, salt spray accelerates corrosion of mounting structures and electrical components. In the Western Ghats and Himalayan foothills, partial shading from surrounding terrain requires careful array configuration to minimize power losses.

Advanced design tools help navigate these complexities. Three-dimensional modeling software simulates shading patterns throughout the year, informing optimal row spacing and orientation. Computational fluid dynamics models predict wind loads on mounting structures, particularly important in cyclone-prone coastal areas. Electrical simulation tools optimize string configurations and inverter sizing to minimize mismatch losses and maximize conversion efficiency.

For utility-scale plants, layout optimization balances multiple objectives. North-south oriented tracker systems maximize annual energy production but require greater land area due to wider row spacing. East-west oriented bifacial systems with vertical panels produce less total energy but generate more during morning and evening peak demand periods, potentially commanding higher tariffs. The optimal approach varies based on land costs, power purchase agreement structures, and grid integration requirements.

In rooftop applications, design optimization navigates different constraints. Available roof area, structural loading capacity, and existing electrical infrastructure all limit design options. Shadow analysis becomes particularly critical in urban environments where neighboring buildings, water tanks, and other structures create complex shading patterns. Microinverters or DC optimizers often prove cost-effective in these conditions, allowing each panel or small group of panels to operate at its maximum power point independently.

Once designed and constructed, solar systems require ongoing performance monitoring to maintain optimal operation. Monitoring systems range from simple energy meters that track total production to sophisticated platforms that analyze performance at the module, string, and inverter levels. These systems typically compare actual production against expected values based on measured irradiance, temperature, and system specifications, flagging underperformance for investigation.

The granularity of monitoring directly impacts optimization potential. String-level monitoring can identify underperforming sections of an array but not specific issues within those sections. Module-level monitoring through power optimizers or microinverters provides the most detailed visibility but at higher cost. The appropriate monitoring architecture depends on system size, performance sensitivity, and maintenance approach.

Data analytics transforms monitoring information into actionable insights. Simple performance ratio calculations compare actual versus theoretical energy production, providing a technology-agnostic metric of system efficiency. More sophisticated approaches use statistical methods to separate expected weather-related variations from actual performance issues, reducing false alarms while catching real problems.

Machine learning algorithms increasingly supplement traditional analytics, detecting subtle patterns that might escape conventional analysis. These systems can identify gradual degradation trends, correlate performance with specific weather conditions, and even predict component failures before they occur. Companies like SenseHawk, SolarSquare, and Prescinto have developed India-specific analytics platforms that account for local conditions like monsoon seasonality and dust accumulation patterns.

Maintenance strategies directly impact long-term performance and financial returns. Reactive maintenance—fixing problems after they occur—minimizes short-term costs but risks extended production losses and secondary damage. Preventive maintenance follows fixed schedules regardless of actual conditions, ensuring reliability but potentially wasting resources on unnecessary interventions. Predictive maintenance uses condition monitoring and failure prediction to perform interventions only when needed, optimizing the cost-benefit balance.

Cleaning represents the most visible and frequent maintenance activity for Indian solar installations. Dust accumulation can reduce power output by 1-2% per week in arid regions during dry seasons, making cleaning frequency a key economic optimization. The decision involves balancing cleaning costs (labor, water, equipment) against energy losses from soiling. Advanced approaches include:

Robotic cleaning systems that reduce water usage and labor costs

Anti-soiling coatings that reduce dust adhesion to panel surfaces

Tilt angle optimization that balances self-cleaning from rainfall against optimal solar angles

Data-driven cleaning schedules based on measured performance impact rather than fixed intervals

Beyond cleaning, comprehensive maintenance programs address electrical and structural components. Thermal imaging identifies hot spots in panels that might indicate internal defects. IV curve tracing diagnoses degradation in solar cells. Torque testing ensures mounting hardware remains secure despite thermal cycling and vibration. Electrical testing verifies insulation integrity and proper grounding, particularly important during India's monsoon season when moisture ingress risks increase.

Vegetation management presents another optimization challenge, particularly in India's high-growth tropical and subtropical regions. Uncontrolled plant growth can shade panels and create fire hazards during dry seasons. Approaches range from conventional mowing to grazing animals (particularly sheep, which rarely damage equipment) to biodegradable ground covers that suppress weeds while preventing erosion.

Security optimization protects against both theft and vandalism, which affect solar installations across India. Physical measures include perimeter fencing, CCTV systems, and security lighting. Electronic monitoring detects unusual activity patterns or unexpected equipment disconnections. Community engagement programs create local stakeholders with interest in protecting nearby installations, particularly effective for rural projects.

Economic modeling provides the framework for all optimization decisions by quantifying costs, benefits, and risks in financial terms. Discounted cash flow analysis remains the foundation, calculating net present value and internal rate of return for different design and operational scenarios. More sophisticated approaches incorporate Monte Carlo simulation to account for uncertainties in solar resource, component degradation, and maintenance costs.

Financial optimization extends beyond technical performance to project structure and financing mechanisms. Capital structure—the balance between equity and debt—significantly impacts returns. Debt tenor (loan duration) and interest rates directly affect project economics, with longer tenors typically improving returns despite higher total interest payments. Tax incentives, accelerated depreciation benefits, and renewable energy certificates provide additional optimization opportunities.

For commercial and industrial installations, integration with energy management systems enables further optimization. Load shifting aligns energy-intensive processes with peak solar production periods. Peak shaving uses solar generation to reduce maximum demand charges. Power factor correction through advanced inverters reduces reactive power charges. These strategies can improve project economics beyond simple energy savings.

Repowering and upgrades offer optimization opportunities as systems age and technology advances. Replacing original inverters with newer, more efficient models can improve conversion efficiency by 2-3%. Adding bifacial modules during panel replacements can increase energy yield by 5-30% depending on ground reflectivity. Retrofitting fixed-tilt systems with single-axis trackers typically boosts production by 15-25%, though economic viability depends on remaining project life and structural compatibility.

Performance guarantees and insurance products provide financial protection against optimization failures. Engineering, procurement, and construction (EPC) contractors typically provide performance ratio guarantees for initial years of operation. Operation and maintenance (O&M) contracts often include availability guarantees with liquidated damages for underperformance. Specialized insurance products cover weather-related production shortfalls, equipment breakdown, and gradual degradation beyond warranted limits.

The optimization landscape continues evolving as India's solar sector matures. Early projects focused primarily on maximizing energy production, but newer installations increasingly optimize for grid integration, time-of-delivery value, and system flexibility. This evolution reflects solar's transition from a niche alternative to a mainstream generation resource with responsibilities beyond simple energy delivery.

Case studies across India demonstrate the impact of comprehensive optimization approaches. The 648 MW Kamuthi Solar Power Project in Tamil Nadu implemented robotic cleaning systems that reduced water consumption by 60% while maintaining higher average performance ratios. The Azure Power installations in Rajasthan pioneered data-driven predictive maintenance that reduced downtime by 35% compared to industry averages. Rooftop installations at Infosys campuses integrated building energy management systems that improved self-consumption rates to over 90%, maximizing economic returns despite modest net metering tariffs.

These examples illustrate that solar performance optimization is not a single technology or technique but an integrated approach spanning technical, operational, and financial dimensions. As India's solar capacity continues its rapid growth, systematic optimization will be essential to maximize the return on these investments and ensure they deliver their full potential for the nation's energy transition.

Power **1.2.02**

Rooftop Solar PV (I & M) - Sun 's Path and Irradiance

Identify various parts of rooftop solar power plant and its uses (On grid, Off grid & Hybrid)

Objectives: At the end of this exercise you shall be able to
- **identify the parts**
- **measure solar intensity.**

Requirements

Tools/Measurements/Equipments

- Solar PV System - on grid - 1 set.
- Solar PV System - off grid - 1 set.
- Solar PV System - hygrid - 1 set.

PROCEDURE

Identify the parts of rooftop solar panel

1 Identify the each parts of a roof top solar PV system
 - ON grid
 - OFF grid
 - Hybrid

Example: PV Module, AJB, MC4 connector, SPO, DC fuse, charge controller, inverter and battery, etc.

2 Record their name one by one in a tabular column.

3 Note down their specification as per the name plate details/ Lasels/ Manuals.

4 Draw the single line diagram (SLO) of the given on-grid, off-grid and hybrid PV system (Fig 1)

Fig 1

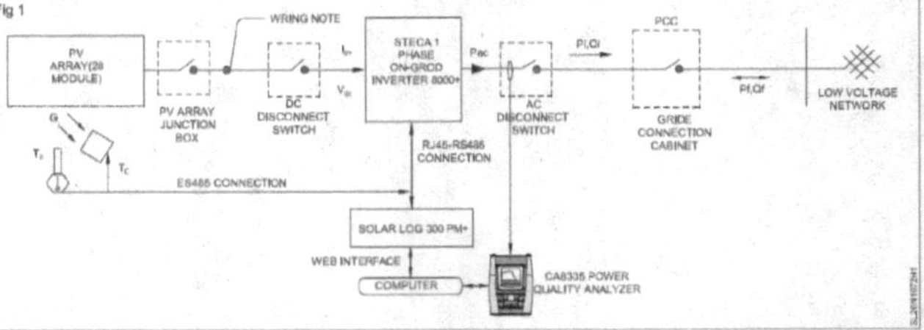

Table

S. No.	Name of parts	Specifications	Remarks

Fig 2

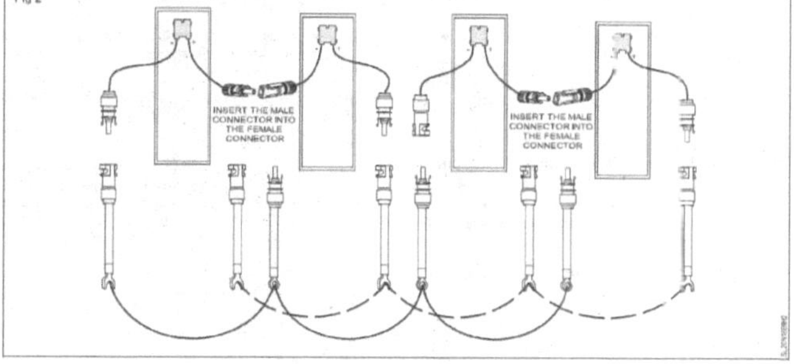

TASK 2 : **Test 4 X 40 W Solar panels in parallel**

1 Replace the 10 W solar panel with 40 W solar panel.

2 Repeat the steps from 2 to 6 of task 1.

3 Record the observations in table 1.

Table 1

Watt/Solar PV module = ...Watts

Load condition	Voltmeter reading (V)	Ammeter reading (A)	Wattmeter reading (W)
No load			
Half load			
Full load			

Observations:

TASK 3: **Recommend places for different components and submit site inspection report of proposed SPV installation**

1 Calculate the space requirement as per the work order.

2 Identify requirement of additional special tools, PPE, material handling equipment or services specific to the work place.

3 Draw a sketch for seating proposed PCU/inverter in safe, shadowed, dry, and adequately ventilated space.

4 Schedule pre requisite activities to be performed to carry out the installation work including clearing old items dumped area on roof with prior discussion with site owner, either through hiring or outsourcing.

5 Estimate additional cost probability specific to this site and inform owner of site as well as company so that the budget can be reset.

6 Submit and store the records including photographs properly in the company.

Consider

Perform Site Assessment so that Components are placed in proper locations where they function without hindrance, Shadows do not fall on the arrays maximizing output and The area remains safe for humans and equipment.

Components of SPV system proposed, layout and flow of energy directions (Fig 1)

Fig 1

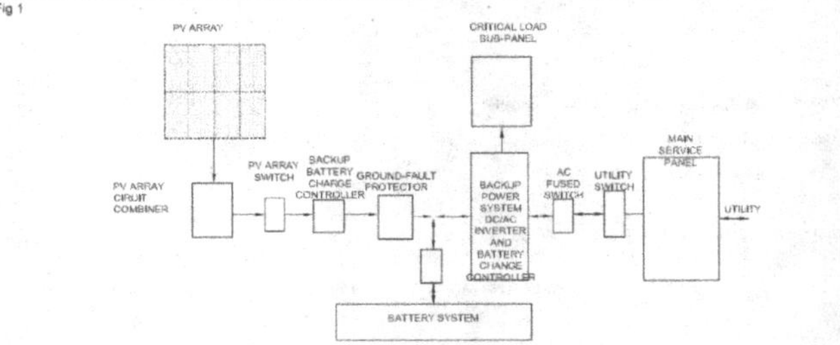

Components of Grid interactive SPV system without battery backup. (Fig 2)

Fig 2

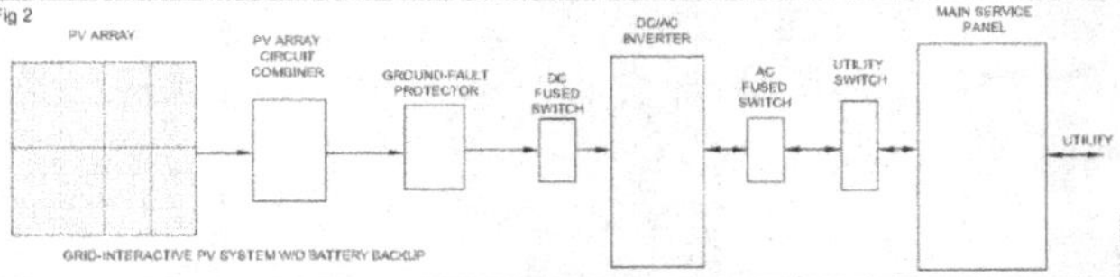

Observation

S. No.	List the observation

- - - - - - - - - - - -

Individual electrical symbols

Symbol	Identification	Explanation
	Transformer	Represents a variety of transformers from liquid filled to dry types. Additional information is normally printed next to symbol indicating winding connections, primary /secondary voltages and KVA or MVA ratings
	Removable or drawout circuit breaker	Normally represents a MV drawout circuit breaker 5kV and above.
	Future removable or drawout circuit breaker position	Represents a structure equipped to accept circuit breaker in the future, commonly known as provisions.
	Non-drawout circuit breaker	Represents a fixed mounted low voltage circuit breaker
	Removable or drawout circuit breaker	Represents a drawout low voltage circuit breaker.
	Disconnect switch	Represents a switch in low or medium/high voltage applications (open position shown)
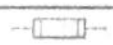	Fuse	Represents low voltage and power fuses.
	Bus duct	Represents low and medium/high voltage bus duct.
	Current transformer	Represents current transformers mounted in assembled equipment. A ratio of 4000A to 5A shown.
	Potential or voltage transformer	Represents potential transformers usually mounted in assembled equipment. A ratio of 480V to 120V shown.
	Ground (earth)	Represents a grounding (earthing) point
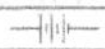	Battery	Represents a battery in an equipment package
	Motor	Represents a motor and is also shown with an "M" inside the circle. Additional motor information is commonly printed next to symbol, such as horsepower, RPM and voltage.
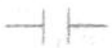	Normally open (NO) contact	Can represent a single contact or single pole switch in the open position for motor control
	Normally closed (NC) contact	Can represent a single contact or single pole switch in the closed position for motor control
	Indicating light	The letter inside circle indicates the color. The color red is indicated.
	Overload relay	Protects a motor should an overload condition develop.
	Capacitor	Represents a variety of capacitors.
	Ammeter	A letter is usually shown to designate the meter type (A = ammeter, V = voltmeter, etc.)
	Instantaneous overcurrent protective relay	The device number designates the relay type (50 = instantaneous overcurrent, 59 = overvoltage, 86 = lockout, etc.)
	Emergency generator	The symbol is frequently shown in conjuction with a transfer switch.

4

Policy and Future Pathways

Government Solar Incentive Architecture

The Indian government has constructed a multifaceted incentive framework to accelerate solar energy adoption across different market segments. This architecture combines financial support, regulatory mechanisms, and institutional structures designed to address specific barriers in the solar value chain. Understanding this complex landscape is essential for stakeholders seeking to navigate India's solar ecosystem effectively.

Central government incentives form the foundation of India's solar support system. The Ministry of New and Renewable Energy (MNRE) serves as the nodal agency, designing and implementing programs that have evolved substantially since the National Solar Mission's launch in 2010. Early incentives focused primarily on capital subsidies—direct financial support reducing the upfront cost of solar installations. These subsidies initially covered up to 30% of benchmark costs for various applications, with higher percentages for special categories like government buildings and social sector institutions.

As the market matured and solar costs declined, the incentive architecture shifted toward more sophisticated mechanisms. Viability Gap Funding (VGF) emerged as a key tool for utility-scale projects, providing capital support to bridge the gap between project cost and revenue potential at competitive tariffs. Unlike simple subsidies, VGF is typically awarded through reverse auctions where developers bid for the minimum support needed to make projects viable at a predetermined tariff.

The Central Financial Assistance (CFA) program represents another evolution in the incentive structure. This program provides targeted support for specific solar applications, with rates and eligibility criteria periodically revised to reflect changing market conditions. Current CFA rates include 30% of benchmark cost for residential rooftop installations, 20% for institutional rooftops, and various rates for specialized applications like solar pumps and off-grid systems.

Tax incentives complement direct financial support in the central government's solar strategy. Accelerated depreciation allows businesses to depreciate 40% of solar asset value in the first year, improving project economics through tax savings. Customs duty exemptions or reductions on specific components reduce system costs, though these have been selectively withdrawn to encourage domestic manufacturing. Tax holidays on solar project income, originally offered under Section 80-IA of the Income Tax Act, provided early-stage support but have been phased out as the sector matured.

The Production Linked Incentive (PLI) scheme represents the newest addition to the central incentive architecture, focusing specifically on manufacturing rather than deployment. Launched in 2021 with an allocation of ₹4,500 crore (approximately $600 million), this program provides financial incentives based on production volume and efficiency for solar manufacturing facilities. The scheme aims to build 10 GW of integrated manufacturing capacity from polysilicon to finished modules, reducing import dependence and creating domestic manufacturing jobs.

Beyond financial incentives, the central government has established enabling mechanisms that indirectly support solar development. Renewable Purchase Obligations (RPOs) require electricity distribution companies and large consumers to procure a specified percentage of their electricity from renewable sources, creating guaranteed demand for solar power. These obligations have been progressively increased, with current targets requiring 21% of electricity consumption from renewable sources by 2022, including a specific solar RPO component.

State government initiatives add another layer to India's solar incentive architecture, with policies varying significantly across states. Leading solar states like Gujarat, Rajasthan, and Karnataka have implemented additional incentives beyond central schemes. These include state-level capital subsidies, exemptions from electricity duty, concessional land allotment, and expedited approval processes through single-window clearance systems.

Gujarat's solar policy exemplifies comprehensive state-level support, combining financial incentives with enabling provisions like banking facilities that allow generators to "store" excess production with the utility for later use. Rajasthan's land policy provides concessional rates for solar projects in designated solar parks while waiving stamp duty on land transactions for renewable energy. Tamil Nadu offers additional generation-based incentives for certain categories of solar installations, supplementing central government support.

Solar parks represent a specialized component of the incentive architecture, addressing land acquisition and infrastructure challenges that often delay project development. The Solar Park Scheme, implemented through the Solar Energy Corporation of India (SECI), provides central financial assistance of ₹25 lakh per megawatt (approximately $33,000/MW) for developing dedicated solar parks with necessary infrastructure including roads, water supply, and evacuation facilities. These parks offer "plug and play" solutions where developers can implement projects without navigating land acquisition and permission processes independently.

Major solar parks like the Bhadla Solar Park in Rajasthan (2,245 MW), Pavagada Solar Park in Karnataka (2,050 MW), and Kurnool Ultra Mega Solar Park in Andhra Pradesh (1,000 MW) demonstrate the effectiveness of this approach. These facilities have achieved some of the world's lowest solar tariffs by reducing development risks and creating economies of scale in infrastructure development. The solar park model continues evolving, with newer parks incorporating storage facilities and exploring hybrid wind-solar configurations to optimize land use and grid integration.

Agricultural solar support programs address the specific needs of India's farming sector while advancing solar adoption. The PM-KUSUM (Pradhan Mantri Kisan Urja Suraksha evam Utthaan Mahabhiyan) scheme represents the most ambitious initiative in this category, with three distinct components:

Component A supports installation of 10,000 MW of decentralized ground-mounted solar plants (500 kW to 2 MW) on barren land, enabling farmers to sell power to distribution companies.

Component B provides subsidies for 20 lakh standalone solar agricultural pumps, replacing diesel pumps and reducing both operating costs and emissions.

Component C supports solarization of 15 lakh existing grid-connected agricultural pumps, allowing farmers to use solar power for irrigation and sell excess generation to the grid.

The PM-KUSUM scheme combines central and state subsidies covering up to 60% of system cost, with farmers contributing the remainder either directly or through bank loans. This program addresses multiple challenges simultaneously: reducing agricultural power subsidies that strain utility finances, decreasing diesel consumption in farming, creating additional income opportunities for farmers, and expanding solar capacity without additional land requirements.

Rooftop solar incentives target the distributed generation segment, which faces distinct barriers from utility-scale development. The Grid Connected Rooftop Solar Programme provides direct financial support through Central Financial Assistance while establishing enabling mechanisms like net metering. The incentive structure differentiates between residential consumers (receiving 40% subsidy for systems up to 3 kW and 20% for larger systems up to 10 kW) and institutional/commercial consumers (who receive accelerated depreciation benefits rather than direct subsidies).

Implementation of rooftop incentives occurs through multiple channels. For residential systems, consumers can apply through designated national or state nodal agencies. Alternatively, they can work with empaneled vendors under the MNRE's SPIN (Solar Photovoltaic Installation) program, who handle subsidy documentation and provide quality-assured installations. Some states have established additional channels through distribution companies, which integrate rooftop solar with existing customer relationships and billing systems.

Off-grid solar support addresses energy access challenges in remote areas beyond the reach of conventional grids. The Off-Grid and Decentralized Solar PV Applications Programme provides financial assistance for solar street lights, study lamps, home lighting systems, and mini-grids serving rural communities. These incentives typically cover 30-90% of benchmark costs depending on the application and beneficiary category, with higher support for special regions including northeastern states, hill states, and island territories.

Financing mechanisms complement direct incentives in India's solar support architecture. The Indian Renewable Energy Development Agency (IREDA), a specialized non-banking financial company under the MNRE, provides loans for renewable energy projects at concessional rates. The World Bank-supported Sustainable Partnership for Rooftop Solar Acceleration in Bharat (SUPRABHA) program offers technical assistance and debt financing for rooftop installations. Green bonds, climate finance, and specialized credit lines from commercial banks provide additional financing options with terms tailored to solar project characteristics.

The implementation infrastructure for these various incentives involves multiple institutions at different levels. The MNRE establishes overall policy direction and program design. SECI handles implementation of large-scale programs including solar park development and utility-scale procurement. State nodal agencies like GEDA (Gujarat), RRECL (Rajasthan), and KREDL (Karnataka) administer state-level programs and often serve as implementation partners for central schemes. Distribution companies increasingly participate in rooftop program implementation, leveraging their customer relationships and technical knowledge of local grid conditions.

Monitoring and verification systems ensure incentives achieve their intended outcomes. The MNRE's Renewable Energy Management Centers track the performance of supported projects. The SPIN portal provides transparency in vendor empanelment and subsidy disbursement for rooftop systems. Third-party evaluations assess program effectiveness and identify improvement opportunities. These accountability mechanisms have evolved in response to early implementation challenges, including delays in subsidy disbursement and quality concerns with some installations.

The incentive architecture continues evolving to address emerging challenges and opportunities. Recent modifications include:

1. Shifting from upfront capital subsidies toward production-based incentives that reward actual generation
2. Introducing quality and performance standards as conditions for incentive eligibility
3. Streamlining application processes through online portals and simplified documentation
4. Integrating storage support with solar incentives to address intermittency challenges
5. Developing specialized programs for new applications like floating solar and agrivoltaics

These refinements reflect the maturing of India's solar market and the changing nature of barriers to adoption. As basic cost challenges diminish with falling technology prices, incentives increasingly target specific market failures, integration challenges, and strategic objectives like manufacturing development.

The effectiveness of this complex incentive architecture varies across market segments. Utility-scale development has exceeded expectations, with competitive auctions driving tariffs below even the most optimistic projections. Rooftop solar has grown more slowly than targeted, hampered by implementation challenges and misaligned distribution company incentives. Agricultural solar shows promising early results but requires careful coordination between energy and agricultural policies. Manufacturing support remains a work in progress, with domestic production still meeting only a fraction of India's equipment needs.

Looking ahead, India's solar incentive architecture faces several challenges. Fiscal constraints may limit the government's ability to provide direct financial support as deployment scales up. Balancing support for deployment versus manufacturing creates policy tensions, particularly regarding import duties and domestic content requirements. Coordination between central and state initiatives remains imperfect, sometimes creating confusion for market participants. Despite these challenges, the overall framework has successfully catalyzed India's solar transformation, creating one of the world's most dynamic markets for solar energy.

Financial Mechanisms for Solar Transition

The financial architecture supporting India's solar transition has evolved from simple government subsidies to a sophisticated ecosystem of funding instruments, investment vehicles, and risk mitigation tools. This evolution reflects both the maturing of solar technology and the financial sector's growing comfort with renewable energy as an asset class. Understanding these mechanisms is essential for stakeholders seeking to participate in India's expanding solar economy.

Project finance forms the foundation of utility-scale solar development in India. Unlike conventional corporate finance where companies fund projects from their balance sheets, project finance creates special purpose vehicles (SPVs) that isolate each solar installation as a distinct financial entity. This structure allows developers to optimize capital allocation, manage project-specific risks, and attract investors interested in particular geographies or project types.

The typical capital structure for utility-scale solar projects in India involves 70-80% debt and 20-30% equity, though this ratio varies based on project characteristics and developer profile. Debt financing comes primarily from commercial banks, non-banking financial companies (NBFCs), and specialized institutions like the Indian Renewable Energy Development Agency (IREDA). Loan tenors typically range from 15-18 years, aligning with the productive life of solar assets and power purchase agreement (PPA) durations.

Interest rates for solar project loans have declined substantially as the sector has matured. Early projects faced rates of 12-14%, reflecting perceived technology and regulatory risks. Current rates for well-structured projects with creditworthy offtakers typically range from 8-10%, with some premium developers accessing even lower rates. This reduction in financing costs has contributed significantly to falling solar tariffs, as interest expenses represent a major component of levelized cost of electricity for capital-intensive renewable projects.

Specialized debt instruments have emerged to address specific needs within the solar financing landscape. Green bonds—debt securities whose proceeds are exclusively applied to environmental projects—have gained traction among larger developers and financial institutions. India's green bond market has grown from virtually nothing in 2015 to cumulative issuance exceeding $10 billion, with renewable energy representing the largest allocation category. Major issuers include ReNew Power, Greenko, Azure Power, and State Bank of India.

Infrastructure debt funds provide another specialized financing channel, particularly for operational solar assets with established performance records. These funds raise capital from institutional investors seeking stable, long-term returns and deploy it into operational infrastructure including solar plants. This "takeout financing" allows developers to recycle capital into new projects while providing investors with lower-risk exposure to renewable assets.

On the equity side, multiple investment vehicles have developed to channel capital into India's solar sector. Private equity firms like Actis, EverSource Capital, and Ayana Renewable Power have raised dedicated funds for renewable energy investments, typically targeting returns of 15-20%. These investors often provide growth capital to mid-sized developers or create platforms that aggregate multiple projects to achieve scale efficiencies.

Infrastructure investment trusts (InvITs) represent an emerging vehicle for solar financing in India. These publicly listed entities own income-generating infrastructure assets and distribute a majority of cash flows to investors, similar to real estate investment trusts in property markets. The Securities and Exchange Board of India (SEBI) has established regulations specifically for InvITs, creating a framework for retail and institutional investors to access renewable energy returns. While still nascent in the solar sector, this mechanism offers promising opportunities for capital recycling and retail investor participation.

Foreign direct investment has played a crucial role in India's solar expansion, bringing both capital and expertise. International developers like SoftBank (through SB Energy), EDF Renewables, Fortum, and ENGIE have established significant presences in the Indian market. Sovereign wealth funds, including Abu Dhabi Investment Authority and Singapore's GIC, have made substantial investments in leading platforms. International pension funds from Canada, including CDPQ and Ontario Teachers' Pension Plan, have allocated capital to Indian renewable assets seeking stable long-term returns.

The financing landscape for distributed solar—including rooftop and small ground-mounted systems—differs substantially from utility-scale projects. These smaller installations typically cannot support the transaction costs of project finance structures, requiring simplified approaches. For commercial and industrial customers, operating leases and equipment financing have gained traction, allowing businesses to adopt solar without large upfront capital commitments.

The OPEX (operational expenditure) model has emerged as a popular financing mechanism for commercial rooftop solar. Under this approach, developers install, own, and operate systems on customer premises, selling power through power purchase agreements at rates below grid tariffs. This arrangement eliminates upfront costs for customers while providing developers with predictable long-term revenue streams. Major players in this segment include CleanMax Solar, Fourth Partner Energy, and Amplus Solar, who have collectively deployed hundreds of megawatts across corporate customers.

For residential rooftop solar, traditional consumer financing mechanisms predominate. Banks including State Bank of India, Bank of Baroda, and Punjab National Bank offer solar loans with terms of 5-7 years and interest rates comparable to home improvement loans. Some states have established dedicated solar financing programs through local cooperative banks or regional rural banks, often with interest subsidies to improve affordability.

Innovative consumer financing approaches are emerging to address affordability barriers. Pay-as-you-save models allow customers to repay solar investments through their utility bills, with monthly payments lower than the electricity savings generated. Solar-specific credit assessment methodologies evaluate repayment capacity based on current electricity expenditure rather than traditional credit metrics, expanding access to financing for customers with limited credit histories.

Risk mitigation instruments play a crucial role in mobilizing capital for solar projects by addressing specific investor concerns. Payment security mechanisms protect against offtaker default risk—a significant concern given the financial challenges facing many distribution companies. These mechanisms include letter of credit requirements, payment security funds, and tripartite agreements involving state governments as backstops for utility payments.

Political risk insurance protects against regulatory changes, contract renegotiation, or other government actions that

action costs for smaller projects, potentially opening new market segments for institutional investment.

Several challenges remain in fully optimizing solar financing in India. Currency risk affects international investors, as solar revenues are denominated in rupees while much of the capital comes from dollar or euro markets. Tenor mismatch persists between the 25+ year operational life of solar assets and the 15-18 year financing typically available. Refinancing markets remain underdeveloped, limiting opportunities to capture improved terms as projects demonstrate performance and risk profiles evolve.

Despite these challenges, the financial architecture supporting India's solar transition has demonstrated remarkable adaptability and innovation. From virtually non-existent a decade ago, solar financing has grown into a sophisticated ecosystem channeling billions of dollars annually into clean energy infrastructure. This financial transformation has been as important as technological advances in driving India's solar revolution, creating pathways for capital to flow from diverse sources into the country's energy transition.

Global Solar Technology Collaborations

India's solar ambitions extend beyond its borders through an expanding network of international collaborations that accelerate technology transfer, mobilize resources, and position the country as a global clean energy leader. These partnerships span bilateral relationships, multilateral initiatives, research collaborations, and commercial joint ventures, creating multiple pathways for knowledge and technology exchange.

The International Solar Alliance (ISA) represents India's most ambitious global solar initiative. Launched jointly by India and France at the 2015 United Nations Climate Change Conference in Paris, the ISA has evolved from concept to treaty-based international organization headquartered in Gurgaon, India. The alliance focuses specifically on accelerating solar deployment in countries between the Tropic of Cancer and Tropic of Capricorn—regions with exceptional solar resources but often limited technical and financial capacity to harness them.

The ISA's operational framework addresses specific barriers to solar adoption through programmatic interventions. These include:

1. Scaling Solar Applications for Agricultural Use: Promoting solar pumps and cold storage to improve agricultural productivity and reduce diesel consumption
2. Affordable Finance at Scale: Aggregating demand across member countries to reduce costs and attract investment
3. Scaling Solar Mini-Grids: Developing standardized approaches to off-grid electrification for remote communities
4. Scaling Solar Rooftop: Sharing best practices in policy, regulation, and business models for distributed generation
5. Solar E-Mobility: Supporting electric vehicle adoption powered by solar energy

The alliance has grown to include 86 signatory countries and works with multilateral development banks, private sector partners, and civil society organizations to implement its programs. Beyond specific interventions, the ISA represents India's emergence as a normative leader in the global energy transition, shaping international cooperation frameworks rather than simply participating in them.

Bilateral technology partnerships complement the ISA's multilateral approach. India-Germany solar cooperation has been particularly productive, with the German development agency GIZ supporting multiple initiatives including the Indo-German Energy Programme and Green Energy Corridors project. These collaborations have facilitated knowledge transfer in grid integration, forecasting systems, and market design—areas where Germany's earlier experience with high renewable penetration provides valuable lessons for India.

The India-Japan Energy Dialogue has established joint working groups on renewable energy and energy efficiency, facilitating technology exchange in areas including floating solar, smart grids, and hydrogen production from renewable sources. Japanese financial institutions including Japan International Cooperation Agency (JICA) and Japan Bank for International Cooperation (JBIC) have provided concessional financing for solar projects, often tied to technology cooperation components.

The United States-India Clean Energy Finance initiative mobilizes public and private capital for clean energy projects, with a particular focus on distributed solar for underserved communities. The U.S.-India Strategic Clean Energy Partnership includes working groups on renewable power and sustainable growth, facilitating regulatory and technical exchanges between the countries' energy agencies. American national laboratories, including the National Renewable Energy Laboratory (NREL), maintain active research partnerships with Indian institutions on solar resource assessment, grid integration, and advanced photovoltaic technologies.

Research and development collaborations form another dimension of India's global solar engagement. The Solar Energy Research Institute for India and the United States (SERI-IUS) represents a flagship partnership between the two countries, focusing on sustainable photovoltaics, multiscale concentrated solar power, and solar energy integration. This virtual joint institute connects researchers across multiple institutions, accelerating innovation in materials, devices, and systems optimized for Indian conditions.

The India-European Union Clean Energy and Climate Partnership facilitates technology transfer and joint research in renewable energy and energy efficiency. Horizon 2020, the EU's research and innovation program, has funded multiple collaborative projects involving Indian and European researchers working on next-generation solar technologies. These include perovskite solar cells, tandem architectures combining multiple materials to exceed single-junction efficiency limits, and building-integrated photovoltaics adapted to diverse architectural contexts.

Technology transfer mechanisms have evolved beyond traditional aid-based approaches to include commercial partnerships, joint ventures, and technology licensing arrangements.

Indian companies have established strategic relationships with international technology providers to manufacture advanced solar components domestically. These arrangements typically include training programs, quality control systems, and ongoing technical support beyond the initial technology transfer.

Adani Solar's partnership with Chinese technology providers enabled establishment of India's first gigawatt-scale integrated solar manufacturing facility producing cells and modules in Gujarat. Vikram Solar's collaboration with Fraunhofer ISE, a leading German research institute, facilitated development of high-efficiency module designs adapted to Indian conditions. These commercial technology transfers complement government-to-government initiatives, creating multiple pathways for knowledge flow.

Manufacturing collaborations have gained particular importance as India seeks to reduce dependence on imported solar equipment. The Production Linked Incentive (PLI) scheme has attracted interest from international manufacturers considering Indian production facilities, potentially accelerating technology transfer in advanced manufacturing processes. Companies including First Solar (USA), LONGi Solar (China), and Meyer Burger (Switzerland) have explored Indian manufacturing partnerships, though geopolitical ten sions and trade policy uncertainties have complicated some (these discussions.

International standards and certification systems prov another framework for technology collaboration. The I national Electrotechnical Commission (IEC) standard solar equipment have been adapted into Indian Star economies to share experiences and best practices. The Global Solar Council provides a forum for industry associations from different countries to coordinate on standards, training, and market development. These platforms allow India to both learn from international experience and share its own innovations with the global community.

India's solar technology collaborations increasingly focus on frontier areas that will shape the next generation of solar development. These include:

1. Integrated solar and storage systems that provide dispatchable power
2. Green hydrogen production using solar electricity for hard-to-decarbonize sectors
3. Building-integrated photovoltaics that incorporate solar generation into architectural elements
4. Agrivoltaics that combine agricultural production with solar generation on the same land
5. Advanced forecasting systems that improve grid integration of variable renewable energy
6. Recycling and circular economy approaches for end-of-life solar equipment

These frontier collaborations position India not merely as a technology recipient but as an innovation partner helping address global energy transition challenges.

The outcomes of these diverse collaborations manifest across India's solar ecosystem. Technical standards and testing protocols developed through international cooperation ensure equipment quality and system performance. Grid integration approaches adapted from international experience enable higher renewable penetration without reliability compromises. Innovative business models inspired by global examples expand access to solar benefits across socioeconomic segments.

Several factors have contributed to the effectiveness of India's international solar collaborations. The country's clear policy commitment to renewable energy creates a stable foundation for long-term partnerships. India's large market attracts international interest from both commercial and governmental entities. The country's strong technical institutions and growing pool of skilled professionals enable productive knowledge exchange rather than one-way technology transfer.

Challenges remain in maximizing the impact of these collaborations. Intellectual property considerations sometimes limit the depth of technology sharing, particularly for cutting-edge innovations. Adaptation requirements for India's unique conditions—including extreme temperatures, dust, grid instability, and diverse applications—complicate direct technology transfer. Coordination across multiple bilateral and multilateral initiatives creates administrative complexity for Indian institutions managing these relationships.

Looking ahead, India's international solar collaborations are evolving from traditional North-South technology transfer toward more balanced partnerships reflecting the country's growing capabilities and ambitions. South-South cooperation has gained importance, with India sharing its experience with other developing countries through the ISA and bilateral relationships. Commercial partnerships increasingly complement government-to-government initiatives, creating market-driven technology exchange alongside policy-driven cooperation.

The future trajectory of these collaborations will likely focus on system integration challenges rather than basic technology transfer. As solar hardware becomes increasingly commoditized, the knowledge frontier shifts toward integration with storage, demand response, electric mobility, and other elements of a comprehensive clean energy ecosystem. India's international partnerships are evolving accordingly, addressing these next-generation challenges through collaborative innovation.

India's global solar collaborations represent more than technical exchanges—they reflect the country's emergence as a pivotal player in the global energy transition. From technology recipient to innovation partner to normative leader, India's journey mirrors the evolution of solar energy itself: from alternative technology to mainstream solution to foundation of a new energy paradigm. These international connections will continue shaping India's solar future while allowing the country to contribute its unique perspective and innovations to the global clean energy community.

As we conclude our exploration of India's solar landscape, several themes emerge that will shape the country's energy future. Solar energy has moved beyond technical feasibility and economic viability to become a strategic imperative—addressing energy security, environmental sustainability, and economic development simultaneously. The technologies, policies, and financial mechanisms discussed throughout this book provide the tools for this transformation, but ultimately its success depends on coordinated action across government, industry, finance, and civil society.

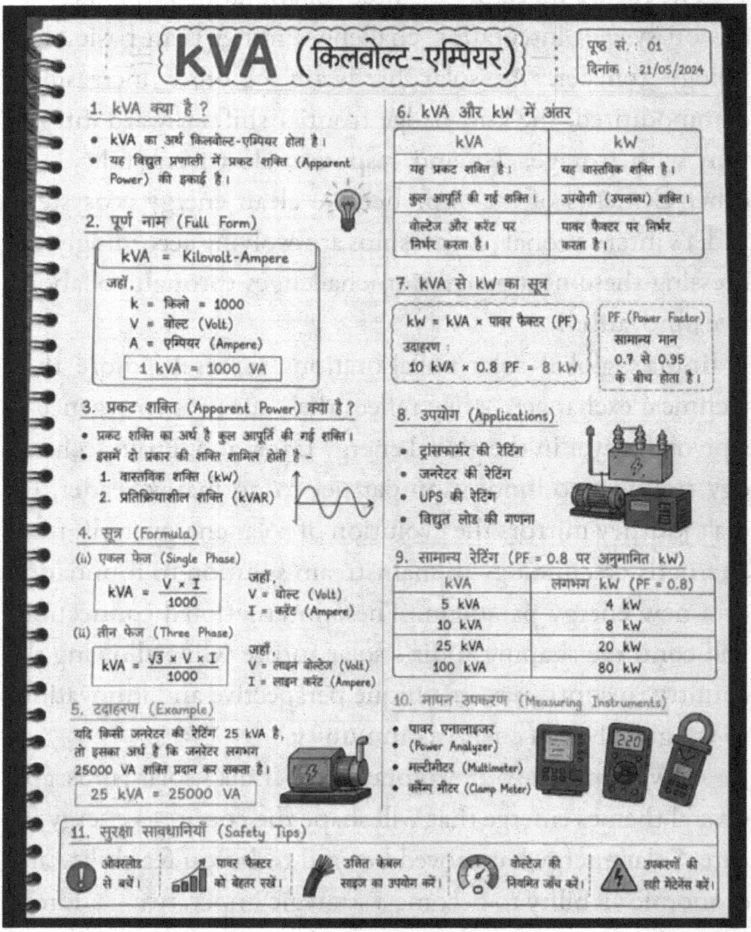

www.ingramcontent.com/pod-product-compliance
Lightning Source LLC
Chambersburg PA
CBHW011349230926
51848CB00003B/522